Flat Earth Meets Tesla:
a Primer in Paired, Inverted Worlds Cosmology

-by Laurel Federbush

copyright 2015 by Laurel Federbush

What's in this book is not your father's physics.

It's not my father's physics, either. (My father is a physicist. He doesn't endorse this book.)

The universe is not as we've been told, and that's a good thing.   It's much simpler.   You don't have to be a genius to understand it, although the research of geniuses like Nikola Tesla does help us to understand some of the processes involved.

(Tesla doesn't endorse this book. He's dead.)

The models of the universe that we learned in school are a relic of the past.   They originated thousands of years ago in the ideas of religions (Vedic Religion, ancient Hinduism, Mithraism, and Zoroastrianism) of Eastern cultures that depended on astrology and worshiped the sun. Some of the sages of ancient India, in particular, were inspired by drug-induced spiritual visions...which could explain a lot about why the their imaginings about outer space were so...what's the word...spacey.

Greek philosophers such as Thales, Anaximander, Pythagoras, and Philolaus were influenced by these Eastern mystical traditions. They even claimed the ideas as their own, as did the mysterious occult figure Hermes Trismegistus–the word "hermetic" comes from his name.

Later, intellectuals like Copernicus and Sir Isaac Newton, students of classical thought–and of the occult--took a liking to the notions as well.   Since intellectuals have a way of seeming to know more than they do, we've adopted their ideas as science.

Since then, we've seen lots of videos and photos that try to present the prevailing theory as fact. Very entertaining.

Now I may not be a rocket scientist

...BUT...

here's what I believe to be a much more accurate depiction of reality, a scenario of the way our cosmos actually works.

I call it Paired, Inverted Worlds Cosmology.

Here goes:

The earth is flat and thin, like a pancake. Above it circle the sun and moon, relatively tiny, and the much-tinier planets and stars. Directly underneath and parallel to it is a second flat earth, the two of them like two stacked pancakes. On the bottom side of the second flat earth is another world, much like ours, with its own sun, moon, planets, and stars--probably people, too--all upside-down from our point of view.

Just to answer a couple of questions before we move on:

1. Doesn't this go against just about every known law of physics?

Yes, but the "laws of physics" that it goes against were invented specifically to prop up the erroneous view of the universe we've been force-fed.   It's like a logic problem: if you make a mistake somewhere along the way, every other conclusion you reach after that is likely to be wrong.
Arbitrarily assuming that the earth is spherical when it isn't has led to one wrong conclusion after another.
We have to get back on the right track.   Admitting that we've been

wrong (yes, it's hard to do) is the first step.

2. Okay, I'll play along. Let's say there are people on the other side of the second flat earth. Wouldn't we have come into contact with these people already, since they're not that far away?

Not necessarily. Since this isn't the model of our cosmos that we've been using, we haven't tried. There may be gravitational, electrical, magnetic, or other physical obstacles preventing contact between the two worlds. However, there have been stories throughout history of people reaching another, similar world underneath ours.

These stories are usually used to support the "hollow earth" model of things.   If my theory is correct, then it could account for some of those stories.   Some of them just sound like legends, others are somewhat more believable.   I'm not sure that any of them are true.   It could just be that, either through intuition or telepathy, we sense that those people are there.

To be clear: that other world isn't Hell, or Heaven, for that matter.

And it isn't some freaky anti-world where each individual person is the mirror image of someone in this world.   There isn't an anti-me and an anti-you.   Let's be real.

It's just a place that's pretty similar to here but slightly different.   Sort of like Canada.

So, to pick up where we left off:

Our sun transmits scalar electromagnetic waves through the paired flat earths.   These waves are received by the moon on the other side.   Likewise, the sun on the other side transmits scalar electromagnetic waves through the paired flat earths, and those waves are received by our moon.   The familiar craters that we see on the face of our moon function more or less as would holes in a telephone receiver.   Each moon is perfectly in resonance with the sun that transmits to it.

A bit about scalar (or longitudinal) electromagnetic waves: Traditional scientists don't generally think these things exist.   They're a Tesla idea.   Scalar electromagnetic waves can penetrate any obstacle, no matter how dense.   They don't lose energy by going long distances. They can affect brain waves.   They are unbounded by time, too, since time is the fourth dimension, and these waves are fifth-dimensional. They can even exist in the absence of matter.   (Imagine weird synthesizer music playing...)

Transmitting scalar waves doesn't take any energy away from the suns, which could go on transmitting indefinitely.   Our sun isn't going to morph into a red giant, or a brown dwarf, or any other fairy-tale creature.

Electromagnetism is a force that attracts and repels at the same time. Not unlike some people.   Have you ever been locked in a relationship with someone you were at once attracted to and repelled by?   Think of each sun-moon pair as being locked together like that.   Each pair is connected by a stationary wave that acts as an invisible leash.

The "leash" goes right through the two earths as if no obstacle were there.   So the motions of each sun is determined by the motions of its corresponding moon on the other side of the earth, and vice versa. The fact that the suns circle at different speeds from the moons causes all of those bodies to move to different circular paths above (or below) their earths at different times of the year, as each body adjusts to the motions of its partner on the other end of the electromagnetic "leash."   Each of the two parallel flat earths acts as a resonator for one of the sun-moon pairs.

All of the energy in our cosmos (which is quite small) results either from these individual stationary waves or from the interaction between the two waves.

The interaction between the pair of scalar electromagnetic waves creates a "Tesla Shield"–a series of globes of energy around us–like bubbles but protective and impenetrable.   We live at the center of these concentric globes.   The stars and planets we see at night are lumination from that phenomenon and exist on the globes, which rotate independently of each other. The ancients who believed that the heavenly bodies rotated on "celestial spheres" were right.

Physicist David Bohm, inventor Arthur M. Young, and others have identified the torus shape (think of a bagel) as a uniquely self-sustaining pattern of energy flow.   The inventor Buckminster Fuller had developed the idea of vector equilibrium, a state where all forces acting on an object are equal.   The torus shape is based on vector equilibrium.

(None of them endorse this book, either...)

The magnetic field is shaped like a dual torus (think of a bagel sliced in half, with the two halves put back together again after you put cream cheese on it).

Through the sliced middle of the "bagel" (where the cream cheese would be) is the flat earth we live on, along with the other flat earth it's coupled with.   Above us arches one half of the magnetic field "bagel," and our counterparts on the far side of the other flat earth get the other half.   Seems fair to me.

So, to sum up:

–Two parallel flat earths sandwiched together.

–The outside face of each has its own sun and moon, probably people, too.
> (Yes, we should probably try to meet them, if possible.   Just to say "hi" or something.)

–Each sun transmits scalar electromagnetic waves through the coupled flat earths to the moon on the other side.

–All of the energy in this relatively tiny cosmos is created by the interaction between these two stationary waves.

–The stationary waves create bubbles of energy around us, on which are luminous points we call planets and stars.

–A magnetic field exists around the paired flat earths in the form of a dual torus.

As far as I know, I'm the first one to put ideas about scalar waves, vector equilibrium, and flat earth together into a theory like this.   Personally, I think they fit together perfectly.   (In my humble opinion.)

This is a basic sketch of how it works.   People who are more knowledgeable than I am about the individual components of this theory can tweak things a bit and fill in the details.

I believe that this cosmological model could help validate the efforts of those committed to developing free and sustainable energy sources.

I also think it's true.

To answer a couple more questions:

1. Conspiracy nut, eh?

No, not particularly. I don't think it's good for one's mental health to focus on who may be conspiring to do what. And as for the minuscule number of people who would have been deliberately deceiving us about the nature of the universe: they probably were or are motivated by concerns about peace, national security, and preserving our freedom. I think people are basically good. The important thing is for us to live our own lives the best we can, while seeking the truth.

2. Okay, even assuming we'd consider your cockamamie idea for a minute: how is it any better as a theory than the usual view of things? Doesn't our current astronomical model (or the many versions of it) explain things pretty well?
No, it doesn't. My theory is way, way better.   Here's why:

–My theory is much simpler.
Occam's Razor says, basically, that the simplest explanation is the most likely to be correct.

--My theory is symmetrical. How cool is that???

--In my theory, everything we see–earth, sun, moon, planets, and stars–functions as part of a dynamic, living machine. Each plays a part. Nothing is superfluous.

–The conventional theory is most complicated, convoluted way possible of explaining what we see around us. You mean there's no way to explain the natural world without having to suppose the existence of a universe zillions of miles across?

--The current model requires matter to behave in ways that we know from our everyday observations–yes, those count–that it doesn't. Put an object on a spinning ball and it goes flying off, no matter how tiny the object or how huge the ball.   But for the usual theory to work, objects have to cling to spinning balls called planets.   Also, inanimate objects don't know how to orbit things. Have you ever seen one do it?

Now my theory has some exotic elements–weird kinds of waves and such–but it doesn't require anything to behave contrary to the way we actually see it behaving.

And my theory has those waves acting in just the ways that Tesla and others have predicted they would. Except, possibly, in this way...

Here's where I part company with some of the others who talk about scalar electromagnetic waves: I think they're inherently a good thing. This waveform seems to be the great creative force in the universe. It envelops and protects its creations. It's everywhere, and can exist even in a vacuum. It's unbounded by time, and can get through any obstacle, probably influence minds...we're probably only touching the surface of what it can do.

I think this waveform is conscious, maybe even consciousness itself. Either God or God's right hand. Consciousness as the creative agent in our cosmos. Not something that would allow itself to be used for weapons of oppression to enslave humanity.

(Not trying to start a new religion–I'm Jewish. I believe that this idea is consistent with existing religions.)

Which doesn't mean that people wouldn't try to use the waveform for sinister purposes, and they may be doing so now. My Jewish tradition teaches that we shouldn't rely on a miracle, so it would certainly be wise to be vigilant. But here's my prediction:

Either it isn't actually possible for humans to use this waveform in a destructive manner, or...

consciousness as manifested in humans will reject any such weapons, or...

the waveform itself will sabotage efforts to use it for malevolent purposes.

How cool is that???

Paired, Inverted Worlds Cosmology.

It could really turn your world upside-down!
But I prefer to see it as right-side-up.

Thank you.

SOME GOOD SOURCES:

ancient origins of "solar system" concept

http://altreligion.about.com/od/alternativereligionsaz/a/zoroastrianism.htm
http://starteachastronomy.com/indian.html
http://users.clas.ufl.edu/ufhatch/HIS-SCI-STUDY-GUIDE/0060_theSunHermeticScientific.html
http://www.crystalinks.com/philolaus.html
http://www.guide-to-astronomy.narod.ru/scientists_and_astronomers/pythagoras.html
http://www.interestingfacts.org/fact/ancient-indian-astronomers
http://zoroastrianheritage.blogspot.com/2011/04/zoroastrian-influence-on-greek.html

flat earth

www.wildheretic.com
William Carpenter: One hundred proofs that the earth is not a globe
Samuel Rowbotham (aka "Parallax"): Zetetic Astronomy: Earth Not a Globe
www.theflatearthsociety.org
YouTube: (Pawel Kolasa) Earth Is Motionless, Copernicus Is Wrong, Parts 1 & 2
YouTube videos on "flat earth" by conspiracydude, freeanergy, jhenningkellogia, Rory Cooper, TheNASAchannel, and others

outer-space fantasies

http://www.apfn.org/apfn/moon.htm
http://books.google.com/books/about/Russia_s_space_hoax.html?id=4hJDAAAAIAAJ
http://www.moonlandinghoax.org/
http://www.nasamoonhoax.com/archives/206
YouTube: (weneverwenttothemoon.com) Fake Moon Landing Totally Exposed

Tesla/scalar waves

http://en.wikipedia.org/wiki/Wardenclyffe_Tower
http://pesn.com/2011/03/26/9501797_Teslas_Scalar_W
aves_Replicated_by_Steve_Jackson/
http://quwave.com/Scalar-Waves.html
http://rationalwiki.org/wiki/Scalar_wave
http://teslascience.org/
http://www.angelfire.com/oz/cv/scalarweapons.html
http://www.bibliotecapleyades.net/ciencia/antigravity
worldgrid/ciencia_antigravityworldgrid09.htm
http://www.cheniere.org/books/analysis/history.htm
http://www.meyl.eu/go/index.php?dir=10_Home&page
=1&sublevel=0
http://www.oocities.org/area51/shadowlands/9654/bea
rden/background.html
http://www.prahlad.org/pub/bearden/scalar_wars.htm
http://www.rense.com/general39/scalarenergy.htm
http://www.rmcybernetics.com/science/physics/electr
omagnetism2_scalar_waves.htm

http://www.teslagenerator.com/the-wardenclyffe-towe
r/
http://www.teslasociety.com/teslatower.htm

torus/vector equilibrium

http://www.arthuryoung.com/astroexc.html
http://bfi.org/
https://prezi.com/ort-oii-ltwk/isaac-newtons-translation-of-the-emerald-tablet/
http://resonance.is/
http://www.botany.hawaii.edu/faculty/wong/BOT135/Lect18.pdf
http://www.cosmometry.net
http://www.mysterium.com/hypercosmic.html
http://www.theatlantic.com/health/archive/2013/12/religion-as-a-product-of-psychotropic-drug-use/282484/
http://www.thrivemovement.com/the_code-fundamental_pattern